Charles Maina

KANG'WE AND THE MONSTERS OF MAIYONI

AF153190

Charles Maina

KANG'WE AND THE MONSTERS OF MAIYONI

The Day the Monsters Lost

JustFiction Edition

Cover image: www.ingimage.com

Publisher:
JustFiction! Edition
is a trademark of
Dodo Books Indian Ocean Ltd. and OmniScriptum S.R.L publishing group

120 High Road, East Finchley, London, N2 9ED, United Kingdom
Str. Armeneasca 28/1, office 1, Chisinau MD-2012, Republic of Moldova, Europe
Printed at: see last page
ISBN: 978-620-0-10532-5

KANG'WE AND THE MONSTERS OF MAIYONI

I

Some time ago, in the land of Maiyoni, there lived an old widow called Wamutheke. Wamutheke lived with her grandchild called Kang'we.

Wamutheke, Kang'we's grandmother was elderly but strong. Her face was lined with wrinkles and when she smiled, a hundred and one lines etched themselves on her skin. She was very hard-working and loved her grandchild very much. She would hurry from the shamba to be with her Kang'we. She would carry sweet potatoes to roast for him, sugarcane and carrots for him to bite into. She would pick wild berries on the way home and when she came home, Kang'we would rush to her and hug her and want to tell her about all he had done that day. She would take him to her bosom and she would feel so good. She loved Kang'we so much.

Kang'we was just five years old. Kang'we was a happy and friendly boy. He loved making friends and would go out and invite everyone to their home. He made friend with boys like him. He made friends with girls. He made friends even with animals and insects! He would follow butterflies as they floated from flower to flower asking them to be his friend. One time he brought a gecko home and as his grandmother was preparing the evening meal, Kang'we brought the gecko out of his pocket to greet his grandmother. The poor lady almost fainted with fright! But Kang'we just laughed.

Kang'we was a playful boy. He loved to play all games and sometimes invented his own games. He would draw a grid on the ground using a stick and try walking on his hands through all the boxes. And he would invite his friends to compete.

His father had made toys for him while his mother played with him. His grandmother always cheered for him when he played with his mother. He invited other children to their compound to play with him. At other times, his father would take him to the village field where he would play with other children.

II

Sadly, Kang'we's parents had died the previous year from a terrible disease that had hit the land. The disease, which was called Mathumbuko had wiped out many people. It was a terrible pandemic.

The disease neither chose the young from the old, the men from the women. Everybody could catch the disease. It started with a cough that would come again and again. Faster and faster and then one would become thinner and thinner. The eyes would sink into their sockets and the hair would fall off. Soon, one was unable to walk and would soon die on their beds. There was much wailing and fear in the land. Many people moved away to areas like Kasondi, Nathiani and Hweti.

Kangwe's grandmother described mathumbuko as a monster that ate everybody in its sight. She said that it was much like the maimo of the evil forest. Actually, some people said that mathumbuko was an ogre that had changed shape and melted into the air, stealing into people's homes and eating them. In his mind, Kang'we saw Mathumbuko creeping on people and jumping onto their necks.

It came into Kang'we's home and struck his parents. It pinned them down and his parents could not move from their beds. They slowly wasted away and one day, Kang'we saw them being buried deep inside the ground just outside their compound.

Kang'we cried and cried and cried. He wondered why it should be his parents but nobody gave him an answer. It was a sad day for Kang'we. He sat at the doorway to his parents' hut, his head in his hands. He could barely summon the strength to hit at the tens of flies that came crawling all over his face, seeming to want to be with him and console him at this time of great sorrow. The burial party stood at the edge of his parent's grave and sang sad songs.

Many people became victims of this disease and it ate and ate as many as it could catch. The disease was seen as a great a monster that had taken residence in Maiyoni.

At night people would speak in whispers lest they be heard by the disease and have it come into their home. It was believed that the monster stalked the land in darkness and that it would emerge from the evil forest at night to stalk the land. Even today, the people of Maiyoni fear darkness and fear walking in the night. As the darkness approaches, everyone runs into the house and lights a lamp to keep the darkness away. That is why every home in Maiyoni has a lamp burning through the night.

III

Kang'we and his grandmother lived at the edge of the evil forest. Being poor and having no one to speak for her, the old lady and her grandson could only get land there. The evil forest was home to many, many ogres called Maimo.

The maimo were ugly monsters. Their skins were scaly like those of crocodiles. An adult Keimo stood at four meters tall. They had one huge eye at the center of the forehead. The yellow eye, with a slit pupil like that of a snake, rolled around in all directions. They had a single nostril below which a wide mouth, with saw-like teeth snapped at everything including flies. The mouth was not unlike that of the Tynasorous Rex. Saliva drooled continously from this mouth.

The maimo stood on a single column of a leg. It extended from a pot-like scally stomach. If you stood close to the evil forest, you could hear the thump, thump, thump of the maimo as they hopped around the forest. At night, you could feel the ground vibrate as the maimo hunted. They never seemed to sleep.

The Maimo ate everything, including each other. If you were a keimo and got hurt, the other maimo would land on you and eat you to the last shred of scaly skin. They particularly loved to eat human flesh. They were on a constant search for humans to eat. Luckily, they did not like venturing beyond they evil forest because they liked the darkness of the evil forest. They also had a fear of people because they thought people were very clever, but they would easily attack the weak and the lonesome.

Another thing about maimo is that they were not very clever. They quickly forgot what they had done a minute ago and even easily forgot their names. But they were very, very strong.

Back to Kang'we.

IV

Kang'we and his grandmother moved to the edge of the evil forest. That is the only place that the other villagers allowed them to stay. They said that Kang'we and his grandmother might also have mathumbuko. The poor old widow had nobody to fight for her. She put her head in her head and wept bitter tears. She asked God to take her and Kang'we and relieve them of their suffering. She wiped off her tears, moved to the edge of the forest and swore to raise Kang'we to be a kind person, not a person to oppress the weak and the old.

Every day, grandma Wamutheke would go to the shamba. The shamba was about a kilometre away. She would leave Kang'we in their small hut and lock the house from outside. She knew that it was dangerous to leave Kang'we playing in the compound as he could be eaten by maimo. She would come in the evening bearing fruits like mangoes and melons, sweet-potatoes and green maize to roast. Kang'we would jump up and down happily, as he gorged on the yellow, juicy mangoes. And this routine went on and on.

Until one day

A kaimo wandered out of the evil forest. A kaimo was a young Keimo and this one was about the age of Kang'we. It thumped about and soon was in Kangwe's compound. Wamutheke had left a sack of castor oil seeds out in the sun to dry.

Castor oil seeds were very important because they were a source of many good things. They were used to produce jelly which was applied on the skin by girls to make it supple and beautiful. It was also a source of fuel for the lamps and for cooking. The hard shells were used to make ornaments and sometimes to make animal feed. So, castor oil seeds were very important to the people of Maiyoni.

The kaimo went to where the castor oil seeds were and began to play with them; he would take them in his three-fingered palms and throw them into the air, and as they fell, throatily sing:

"The hail is faaalling, faaalling, faaalling!"

Kang'we heard the kaimo playing with his grandmother's seeds.

"Who is that playing with my grandmother's seeds?" he shouted from within the house.

"It is me", the kaimo growled." And who are you and where are you?" the kaimo looked around but could not see Kang'we.

"My name is Kang'we and I am inside the house. Come to the door and let me out!" Kang'we had been warned by his grandmother not to talk to strangers but he had forgotten this.

The Kaimo approached the house but he did not know what a door was.

"What is a door?", asked the kaimo, his single eye rolling this way and that, searching for a door.

Kang'we went to where the door was and spoke behind it.

"What kind of a person are you? Don't you know a door? Here, I am knocking it now. This is the door."

Maimo have a very sharp hearing and the kaimo quickly hopped to where the door was. But he did not know how a door was opened.

"There is a latch outside. You slide it to the left and push, the door will be opened", explained Kang'we. However, the kaimo heard only the last part. He pushed the door in with his scaly arm and the door broke in two.

"Waah! What have you done you... you, my God! You are ugly! And you have a very bad smell! Who are you?" Kang'we was shocked when he saw the ugly kaimo.

"My name is...my name is... I have forgotten my name," the kaimo tried scratching his tiny head to remember his name but it was in vain.

"You are a keimo!," exclaimed Kang'we. "You are supposed to be a very bad thing and you eat people and you..."

The kaimo did not know what to say. He however wanted very much to be a friend to Kang'we because he did not have any friend. It was so lonely and dark in the evil forest.

His parents were always away. Kang'we looked at him and although the kaimo did not know what kindness was, he felt that the way Kang'we looked at him was warm and friendly.

"I have never eaten people. I have eaten flies, snails and bats but I have never eaten people. I just want us to be friends," the kaimo pleaded.

Kang'we could not resist that and he agreed to be friends with the kaimo.

"OK, I will call you…what…Kalondu…is that ok?" asked Kang'we.

"That is a nice name…and what is your name?" asked the kaimo now named Kalondu. He was grinning from one end of the head to the other end. His sharp teeth glistened in the morning light, covered as they were in gluey green droll. His eye widened even more in his happiness, the dark slit of his pupil narrowing even more. Kang'we had never seen such an ugly creature but he was now a friend.

"I already told you my name is Kang'we."

V

Kang'we led his new friend into the yard. Kalondu came hopping after him, his evil smell flowing after him, attracting tens of flies. Kang'we did not mind the little ogre. He had completely forgotten his grandmother's admonitions against strangers and what's more, she had countless times told him of the evil ogres lurking in the evil forest. But all Kang'we wanted was friendship with everyone, to play with friends, to laugh and have fun.

"What game shall we play?" asked Kang'we.

The little monster did not know of any game. In the evil forest, the only thing they did was eat, eat leaves, insects, beetles, eat anything that can be chewed. Kang'we tried teaching Kalondu football, but Kalondu could not kick the little ball made of rags, since he had only one leg. Kalondu could also not cartwheel or do hopscotch.

They eventually decided to sit down and talk. They went and sat down under the big mugumo tree growing in the compound. The huge tree had a huge roomy cavity at its bottom and they got in there. If one had heard them hidden from view, it would have been difficult to know which child was human and which child was an ogre. They talked about frogs and sliding down the slope of a muddy hill. They talked about their grandmothers. Kalondu did not know the difference between a father, mother or grandmother, but he pretended for his new friend.

Kang'we told Kalondu about his father. How he carried Kang'we on his shoulders and ran around with him as Kang'we screamed with excitement and fear. And his father would take him along to herd the goats. And how his mother would come to where they were herding the goats and bring them sour milk and sweet potatoes and…

Kalondu was snoring, he had fallen asleep as Kang'we dived into his sad and happy memories. He had a good look at his new friend as he snored away, His open mouth heaved, harrumphed and hugghhreeed and green saliva boiled and frothed out of it as the little monster dreamt away of whatever little monsters dreamt about. Kang'we could see sharp teeth that lined both sides of the mouth.

It was then that his grandmother's warning came to his head.

He remembered what Wamutheke had told him about the ogres of the evil forest.

He jumped up in fear, his eyes wide open in growing terror at the danger he was in. He saw himself being chewed to jelly between those jaws and being swallowed down that scaly throat. Kang'we began shaking like a leaf. He realised the danger he was in if the little ogre got up hungry. It would definitely have forgotten his name, let alone his friendship. A cold trickle of fear slithered down his spine and his teeth started chattering in fright.

He got up and ran into the house. He tried to get back the broken door into one piece but it wouldn't hold. Shaking with fear he let it be and quickly slipped under the bed. He could hear his heart hammering in his throat. He got into a sack of castor-oil seeds and covered his head with tinder, fine wood-shavings, which his grandmother kept for lighting the fire. He thought that he was now well hidden from Kalondu. He prayed and prayed that his grandmother would come home early.

Oh no! He immediately realised that the little ogre might eat his grandmother if she came home early.

'Oh my God! Please do not let my grandmother come home now!' prayed Kang'we.

Kang'we started crying in terror and remorse. He did not know what to do. He promised God that he would henceforth be a good boy if only he got the little ogre out of their compound.

'Please God, get Kalondu out of our compound and I will...', he searched in his mind for a good promise '...I will...I will give you half of my sweet-potato when my grandmother comes home but please let her not come now!'

At that particular moment, Kalondu was roaring awake.

Ogres do not wake up like you and I.

When they are waking up they make a terrible roar and storm out of their sleep, starting out in a scramble, hopping madly out only to stop and try to recollect where they were. And in their sleep-waking pandemonium, they would knock into others and run into objects be they trees or rocks. So, their waking up is always accompanied by fights with other ogres or injury to faces and bodies. And since, they had only one leg; the scramble is awkward and ridiculous, always ending in a headlong plunge.

Kalondu let out a terrible ear-shattering roar as he started from sleep. He burst up, trying to scramble out of the mugumo cavity. Kang'we could distinctly hear him and he shuddered further – if that were possible – into the sack. Kalondu knocked his tiny ogre-head into the sides of the cavity and yelped out a howl of pain. He got up again and threw himself into the triangle of light, which saw him land squarely in the middle of the compound.

He could not remember where he was. He scanned his surroundings with his yellow eye. The compound shimmered in the afternoon sun. He could not stand the glare of the sunlight and he hopped and thumped his way over the fence, across the field and into the evil forest.

VI

When Wamutheke came in the evening, she was horrified at the broken door. The door could not hold and it was in two pieces. She could see that the force that broke the door was so immense that it was as if it had been sawed in two. She found Kang'we sitting at the hearth . He seemed to have been crying.

Wamutheke rushed to her dear grandchild and held him to herself. She wanted to know what had happened.

'A great bull came and knocked down the door,' said Kang'we. 'I was so afraid that I hid under the bed.'

'Whose bull, my husband?' Wamutheke asked. She called Kang'we her husband because he was named after her long-dead husband Kang'we wa Mutuku.

'I don't know. It was chasing another bull.'

Wamutheke knew that Kang'we was not telling the truth because he kept biting his right forefinger, something he did when he was uncomfortable with what he was saying.

Kang'we was afraid to tell his grandmother the truth. He feared that she would scold him for making friends with an ogre and playing with it. It felt terrible in his heart to lie to his grandmother, but he did not know what else to do.

Soon, his grandmother brought out the sweet-potatoes and put them into the hot ashes to bake. Kang'we would soon be tucking into them, their starchy fullness dancing his cheeks up and down, in and out and chasing them down a tiny happy throat with sour milk sweetened with honey.

She also brought out the paw-paws and the bananas. She prepared the fruits for him.

Kang'we sank his teeth into their yellow goodness as huge droplets of juice fell on his chest, rivulets of sweet syrup dripped down between his fingers. He licked his hands down to the elbow. He soon forgot all his troubles as he took a satisfied nap, his grandmother rocking him back and forth.

Kalondu and all the monsters of the world were soon forgotten in the sweet lullaby of his grandmother.

Kang'we slept. Soon he was swimming in the river that ran down the ravine and into the evil forest. He was with another boy, whom he seemed to know well, but did not know from where. Suddenly, the boy looked exactly like him and he called him Kang'we. The boy shrieked in joy and took Kang'we by the hand, drawing him further into the river. It was getting deeper and deeper. Kang'we started getting worried as it also got darker and darker. He screamed for the second Kang'we to let go of his hand. He was sinking into the river. He looked around him for a reed to hold onto. There was none. The second Kang'we multiplied into two Kang'wes, then four, then eight. Then there were many. They were all holding his hand tightly.

He screamed for them to let go. They did not seem to hear him as the river got louder and louder. And it got darker and darker. He screamed more. All the Kang'wes suddenly turned into little Kalondus, roaring at him. Laughing at him in their terrible ogre-laugh. Kang'we screamed louder and louder! He suddenly lost his voice and soon his mouth looked like the ogre's! He screamed in an ogre manner but no sound came. He was turning into an ogre!! He got under the water and could not breathe. He could not...breathe...he...could ...not...bre..

He suddenly woke up. He was sweating and shaking like a leaf. His grandmother was at the side of his bed, watching him.

Wamutheke did not know what to make of Kang'we's obvious nightmare. This was the most violent one Kang'we had ever had. He had thrashed and kicked his feet, cycling and stomping. He held tightly to her clothes, grunting heavily and the balls of his eyes showing their whites in a seeming seizure. And he sweated profusely.

She tried to wake him from the terrible nightmare but it was futile. Her dear 'husband' was coming down with a fever. She could not help but shed tears of fear and compassion. She could not bear losing Kang'we to a fever. He was her reason for living. If he died, she was sure she would also die shortly. She went down on her knees and prayed to God to spare her dear grandson.

Kang'we was crying. Great heaving sobs. He cried because of the ogres that were trying to drown him. Why would they? Did he do them any wrong?

He cried because he had to lie to his loving grandmother. He cried because he was afraid that Kalondu might come again. But he consoled himself that ogres did not remember a thing. And Kalondu would not remember him. The little ogre would not be able to find him again.

With that self-consolation, Kang'we fell into a deep sleep, devoid of dreams, nightmares, multiplying ogres and other terrible monsters that rose in the dark ; from the bed to torment little boys and girls.

VII

When he woke up, it was already day-light. His grandmother had already left. He found that she had left a gourd of sour porridge and another of sour milk. There was also a bowl full of sweet-potatoes and yams. He looked at the door. His grandmother had tried to tie the two halves of the door together, using twine and sisal cords. He sat down at the hearth and ate his break-fast. He hoped that Kalondu would not come. He peeped out to see whether his grandmother had left the castor-oil seeds in the court-yard to dry in the sun.

Ohh my! There they were!

But he knew that the little ogre would not come. It had forgotten everything about yesterday, Kang'we, the time they spent together. Kalondu was now gone! Kang'we felt so happy.

A salve of peace and well-being descended upon him. He felt even hungrier although he had just eaten. He ate some more. He played in the house, jumping onto his bed and pretending that it was a boat. He rowed back and forth. He set about looking for cockroaches and ants, pretending that he was a big ogre that was going to chew them alive.

He found one cockroach on the rack of utensils. He caught it and tore its legs off so that it would not run away. He was now Mathunya, the big bad ogre.

'Do you know me?' he bellowed, rolling his one eye and closing the other like a big evil one-eyed monster. 'Do you know me? Mathunya wa Ithunya who drinks your body and eats your blood!! ?' He hopped on one foot and bellowed at the poor cockroach.

Time passed and Kang'we never noticed as he played all the games he played when he was left in the house by his grandmother.

Suddenly there was a rustle at the door. Somebody was trying to open the door! Kang'we was near the rack of utensils and he jumped right under it, his heart thumping in his chest. The ropes tying the door together were loosened and the weakened door gave way. Somebody entered with a huge load of vines covering the head and face. Kang'we wanted to scream but no sound came from his throat. As the vines were put on the floor, Kang'we almost fainted with relief. It was his grandmother.

The grandmother looked at her beloved ward and knew that something was wrong. He was wild-eyed, shaking like a leaf. She put her finger under his neck and felt his temperature. She peeped under his eyelids and looked at the palms of his hands. Everything seemed ok. But since the previous day, Kang'we had been behaving in a queer manner.

'My husband, what is wrong?' she asked? But Kang'we did not muster the strength to tell her that he had seen and played with an ogre. That he had actually liked playing with the ogre.

VIII

By the next morning, Kang'we's fears had subsided. He had slept well and Kalondu seemed just a curious incident to be forgotten. As usual his grandmother had locked him in. She had repaired the door with the help of Matwele, her neighbour. Matwele had looked at the broken door and muttered that whatever had broken the door into two must have been really strong. Kang'we quickly found something to do...

It was not until afternoon that the sound of falling castor-oil seeds was heard.

'The hail is faaling...faaaaling...faaaling', Kang'we heard. Kalondu was out there!

The door was locked and he was safe...he checked it again and knew that nobody could get past that door easily...

Kang'we's tiny heart was pounding in his throat! The grating voice of Kalondu dried the saliva in Kang'we's mouth and he felt his ears heat up. A lump of fear and loathing stuck in his throat. He stuck his fingers into his ears to block out Kalondu's sing-song but he could still hear Kalondu as though he was shouting from a pot...

Then he remembered Kalondu was playing with his grandmother's castor-oil seeds. *His grandmother's seeds!* A well-spring of rage rose up from somewhere in his belly and quickly uncoiled to his aching throat.

'WHO IS THAT PLAYING WITH MY GRANDMOTHER'S SEEDS!' Kang'we shouted.

'It is me' came the answer.

'Kalondu?

'Yes'

The young ogre remembered to hop over to the door and knocked it out.

'Now what have you done!?'Kang'we was aghast. The door once again lay in two pieces. Kalondu was so sorry about what he had done. He implored Kang'we to let him stay and let him play with him.

Kang'we knew that his grandmother would soon be home. He did not want her to come and find Kalondu in their home. Kang'we feared that the ogre might even eat his beloved grandmother.

'No,no,no' he firmly said. ' It is almost evening and your parents are looking for you...!'. But Kalondu would not leave. He kept hopping about on his one leg, dancing around Kang'we, pleading with him to play. And anxiety kept building in Kang'we as he looked at the footpath along which his grandmother would come. He was scared that he might put his grandmother in danger.

'Ok! If I give you my finger to go and play with, will you leave right now?' asked Kang'we.

Instead of the little finger Kang'we pointed at him, Kalondu sliced the arm of Kang'we in one deft swipe. His canine teeth were so fast that initially, Kang'we felt no pain.

Then the pain hit him and Kang'we let out a howl of a scream. His scream was so loud that bats in the caves of the evil forest fell in a daze. His scream hit the heavens and bounced off the earth like bolts of fire. Wamutheke, in her garden felt a jolt of apprehension rip through her heart. She knew something was wrong at home. She gathered her tools and quickly headed home.

Kalondu had quickly chewed Kangwe's arm as if it had no bones. He gulped it down his scaly throat and licked his mouth all in one go. He liked it. But Kangwe's scream scared Kalondu horribly as he had never heard such a howl. He hopped away in fear and jumped into the evil forest.

IX

When Wamutheke came home, she found Kang'we in a faint. Blood was all over the place. The old lady felt her weak old heart jump to her throat. She felt her skin crawl away, terrified that her precious Kang'we was dead. She quickcly put her basket down and rushed to Kang'we where he lay at the fireplace. He lay on his left side, She turned him round. Then she saw it.

Kang'we's left arm was sliced off at the shoulder. The shreds of skin that had fallen between the fangs of the beast that had bitten had her dear grandson unsuccessfully attempted to cover the hole that was the shoulder socket, shyly shrinking back up the shoulder blades as if they were embarrassed at their ineffectual attempt.

Then the stink hit her nose. There was only one creature that had that awful smell. The evil ogre. It was the smell that was a cross between rotting flesh, bowel gas and the smell from a garbage dump. There was an ogre in the house! Another wave of terror hit Wamutheke. She quickly looked around the house, under the bed, under the food store and under the utensil rack. No ogre.

Rivulets of sweat were now cascading down the aged lines of her face. Her throat was parched and dry with her effort to find the ogre that had killed her dear Kang'we.

Then she saw the broken door. Then it all came to her. An ogre had broken in the day before yesterday but it seemed to have left without hurting her boy. It is obvious Kang'we never really had a glimpse of it, since he was unable to describe it and thought it was just a great bull. Alas.

There was one thing she was unable to explain though. Why did the ogre or ogres just eat Kang'we's arm? What she knew of ogres is that they ate one whole. They did not know how to eat bit by bit...

As these thoughts were swirling in her head, it hit her like a hammer that her beloved was dead! Kang'we was dead.

She let out a howl of pain, a scream of anguish that repeated itself again and again at the unfairness of her world.

The world that had punished her for the death of her children. The world that declared her poor and unwanted, and pushed them to the edge of the evil forest. The world that felt nothing for people like her. She screamed and screamed for Kang'we, for children like him, sweet souls that the world punished for no blame of their own.

Her screams descended into heaving sobs. She beat the bare earth with her hands, crying the cry of the bereaved, the doomed. She crawled to where Kangwe lay and took him into her arms, his bloodied shirt transferring some of its scarlet load to her skirt. She cried into his face, warm bitter tears and mucus falling on his cherubic face. It is then that Kang'we opened his eyes.

At first, the grief-stricken grandmother did not notice that Kangwe had opened his eyes. He regarded his grandmother with sadness and remorse and called out,

'Cucu, I am thirsty.' He repeated the words a little louder.

Wamutheke was so happily surprised she almost dropped Kang'we into the fireplace.

'At once, my husband!' She rushed to the food store and pulled out the gourd of sour milk, gave it a few shakes and pour out a glass for her beloved husband. She picked him up and went to the bed and sat him on her lap. She fed him the milk in moderate sips. After Kang'we finished the milk, Wamutheke let him lie on the bed while she went out for a few things.

One, she needed a few herbs.

She needed the *mukwera* which she would grind into a paste and pour into the wound. This would kill the pain that she knew would hit with a vengeance and throb like a drum throughout the night. She needed the bark of the *mutindira*. This would kill any tiny worms that may be attracted by the blood and raw tissue to come and feast. Lastly she needed the flowers of the *muthithi* which had soporific effects. She needed her beloved grandson to sleep the sleep of the dead so as to forget all the pain and horror he must have gone through. The only place she could get these were inside the evil forest. What had to be done had to be done.

Secondly, she needed to figure out how to secure her door-less house. If they were to sleep peacefully and avoid unwelcome visitors, she had to find something to secure her doorway. She would need to go to neighbour's farms, looking for mislaid logs and pieces of wood.

It is not to steal, if it is to heal.

XI

When she went back home with what she needed, she found Kang'we wailing in pain. She quickly ground the leaves of the *mukwera,* adding a little water and her spit as she prayed incessantly for her dear grandson. When the green paste was ready, she poured it into the gaping hole of Kang'we's shoulder. In spite of his exhaustion, Kangwe let out another howl as the medicine sizzled and steamed the pain away.

Wamutheke put the *mutindira* and the *muthithi* to the boil. Soon they would be ready. Then she took her dear grandson into her arms, wrapping him in blankets and rocked him quietly as he sobbed his pain away.

'My husband,' said Wamutheke, 'who did this to you? I will not be angry, just tell me who.'

'It is Kalondu,' replied Kang'we weakly.

This was a fresh puzzle for Wamutheke. She had expected him to tell her that it was a keimo. Kalondu then must be a human being from around. What about the smell of the maimo that was all over the house? But again, why would a keimo just eat an arm? A keimo would not have left Kangwe alive, it would have eaten him completely.

'What does Kalondu look like? I will look for him and make sure he does not hurt you again. What does he look like?

'He is ugly,' said Kangwe, 'He has one eye and a big mouth.'

Wamutheke now knew that Kalondu was a keimo. She felt terror leap again from the bottom of her belly, grasp her heart in a cold grip and kick her lungs out of place.

A keimo had been in the house and had eaten a bit of her beloved Kang'we! For some reason, it had mercifully not eaten him wholly.

Another puzzle was why Kang'we called it by a name. Maimo were unnamed. They could only be identified by their worst predilections. So a keimo could not introduce itself to Kang'we with a name.

This was so mysterious. She did not want to bother her Kang'we with any more questions. She secured the door the best way she could, prepared the rest of the herbs and prepared sweet potatoes for her grandchild to take with sweetened sour milk. Then she settled down to watch over him. She knew she would not sleep, and the following day would be a long one.

XII

The following day found Wamutheke at the Area's Voice compound. It was a long walk from the edge of the evil forest, a whole four-hour brisk walk. Wamutheke carried two pumpkins and sweet-potatoes in a sack, as it was known that nobody visited the Voice with nothing in their hands. She had risen well before sunrise for this visit.

The Voice's compound was bigger than her farm. The visitors' hut was bigger than her own compound and the main house could take in ten of her hut. It had beautiful hedges and fruit trees and strange birds that fanned out their tails like her winnower.

The paths that led to the various houses were inlaid with beautiful coloured stones that snaked around green, downy grass. Wamutheke thought the Area Voice was the richest man in the world.

His wife, Ms Kimeria came over and greeted Wamutheke. She was very fat. Her round shoulders cascaded down to massive arms that were segmented into sections. Her torso also was segmented into massive folds of flesh that were supported by the thinnest of legs.

She had no neck to speak of while her round cheeks shorn like oiled melons. Of those parts that were visible, her skin had a sickly yellow colour to it with green veins visible below it,

Wamutheke offered her hand in greeting but Madam Kimeria did not see it. She was more interested in the sack that Wamutheke carried. She took it from Wamutheke, quickly peeked inside and called one of the servants to come for it. She was satisfied with what Wamutheke had brought. Then she turned to Wamutheke, almost angrily…

'What do you want?' enquired Ms Kimeria, menace and contempt in her shrill voice. Her voice was so cold and hateful that you could feel the temperature go down a few degrees. Wamutheke clutched her shawl closer.

'I must see Hon Kimeria…'

'Didn't we tell you not to ever come here?' Madam Kimeria interrupted Wamutheke.

'Yes, but…'

'But what?' barked the fat lady. 'You know very well that your children died from that terrible disease, mathuwhatever, then you come here…what do you want? Do you want to kill us? Have you been sent by Baba Saroni's enemies, so that you kill us all and take his seat?'

By now, Madam Kimeria was almost shouting.

Wamutheke burst out in tears. She could not take the mention of her dear dead children in such a tone. Madam Kimeria's words cut deep into her heart, rending her soul in twain; the pain of loss and the dull throb of fear hit her knees and she collapsed into herself.

Ms Kimeria turned round and walked into her house. She went into her husband's bedroom, where he sat beside their massive bed, wiping the last shreds of mutton and eggs from his breakfast tray.

If the female Kimeria was fat, the male Kimeria was the exact opposite.

He was a reed of a man. He was thin, tall and black. He had a huge flat forehead that ended in bushy eye-brows that looked wild and rebellious and were separated by a narrow bridge that cascaded into a beaked-nose.

Below them were massive owl-eyes whose tiny pupils swam in blood-streaked whites. He had the look of an Angry Bird...His mouth was however the most disconcerting feature. When he closed it, it was a tiny upturned line, lipless and sulky. But the moment he opened it, to eat, laugh or talk, it became a massive cavern and stretched from jowl to jowl. The villagers said the mouth was made for gobbling up everything.

It is true. Mr. Kimeria was greedy. He ate everything and anything. And he took away anything that did not have roots from the villagers if they were not vigilant enough.

He regarded his wife ruefully.

'What is it now?' he asked.

'It is that Wamutheke...she is at the gate.'

'What?! What does she want?' his huge mouth fell apart, pieces of flesh hanging from sharp incisors.

'I do not know. I kindly asked her what she wanted and she started crying,' said Ms Kimeria.

'Heh? Could it be that grandson of hers? Maybe he has also died from their disease...I have nothing, nothing to give her. Did she come with anything?' His eyes shone in angry doubt but greedy hope.

'Yes. She brought two pumpkins...'

'That is not enough to disturb my morning!' bellowed the Area Voice. 'Have you washed your hands or are you going to bring their disease into my house?'

'I did not touch her...'

'You touched her sack...go wash your hands now!'

'Will you go and see her. It is you she wants to see!'

'Oooh God! What did I do to deserve this?' he asked rolling up his blood eyes at heaven. 'Did you tell her I was in?'

'I did not tell her you were *not* in...'replied Ms Kimeria.

'You fool! Why didn't you tell her I was out inspecting development projects? Now she is going to infect me with all the diseases they carry! Lord Jesus, come now and surround me with your blood!'

XIII

Kimeria got off his bedside chair. He was still in his pyjamas. He put on his slippers and prepared to go out and meet Wamutheke. He went into the bathroom and wiped his mouth. He applied a little salt to his lips to get rid of any fat, and look as dry lipped and hungry as possible. He also sucked in his stomach.

Lastly, he looked at himself in the mirror and practised a smile, his meet-the-people smile.

When Kimeria came out of the house, Wamutheke had composed herself. It was now almost nine in the morning. She quickly went through her mind what she wanted to request. Everyone new Hon Kimeria was a busy man, so you went directly to the point.

Kimeria came to Wamutheke all smiling and with his arms wide open. He avoided her hand and took her in a bear hug. Wamutheke avoided looking at his face, they said his eyes saw into one's soul.

'Now, what is it my dear Wamutheke? What blessing has brought you to my homestead so early?'

Wamutheke could not hold herself any longer and she burst into fresh sobbing, heavy and heaving,

'There, there...it cannot be so bad...'Kimeria consoled her as he took his seat five feet across from her.

'What is it?' asked Kimeria.

'Monsters are eating my child! They have invaded our home and they will finish us!' Wamutheke sobbed.

Kimeria had prepared for a plea for financial assistance, money to buy food or to help put up a new roof. This was an absurd statement. He inquired from the air around him what was wrong with people!

'What monsters? Wamutheke, monsters come in many shapes. What monsters?'

'Maimo! Ogres...'

Kimeria cut her short with a sarcastic laugh.

'Wamutheke, Maimo no longer exist in the modern society. They are only found in the imagination of old women like yourself'

Wamutheke pleaded with Kimeria to listen to her, She explained that she and her grandson had been banished to the edge of the evil forest after her children died from Maathumbuko. She reminded him that he was one of the people that approved that she be banished there to avoid any father spread of the disease. Here she broke again into fresh wailing.

'The maimo are very much alive in the evil forest. They just ate half of my grandson!'

Kimeria did not know what to make of this wild ogre story. He had never gone near the evil forest. He had grown up in the city and only came to Ukia to take up his father's seat as the Area's Voice in the city. He had heard of the existence of Maimo from old folks but he thought they were just fictitious characters created to scare children.

Now here was Wamutheke claiming to have seen one eat her grandson. He was not going to entertain this nonsense. He knew she wanted another house, or farm or whatever. He was not going to entertain greed. People were not going to hide behind fantastical tales to perpetuate greed. That was it. She was being greedy, and it's not like she had not lived long enough to know the dangers of greed. What is it with people? Can't a person just know that there is a limit to what you can have? Greed is the bane of man, and women, no matter their...

'Please mheshimiwa...please...' Wamutheke, kneeling before him, jolted him from his thoughts as she grabbed his hand. He jerked away from her, horrified by her touch.

'Ok...ok', Kimeria comforted her. ' We are going to get you another farm and another house, bigger and better away from the evil forest.' His mouth spread out in a massive smile that barely went beyond his nose. His eyes shone with a new intensity.

'This is what we are going to do. I will call the Area Lands Allocator and get him to give you another piece of land where we, as the Area Improvement Committee Initiative, will build you a house, away from the maimo. But you know, I cannot go to the Area Lands Allocator just like this.

Go and get together something that I can add to my own to visit him.'

Wamutheke did not know what to say. She knew she had nothing. But she also wanted another house, away from the evil forest to save her dear Kang'we.

'I..I...I do not have much. I am a poor widow...' Wamutheke said.

'You must have something. Go and get together everything in your farm. Sweet potatoes, pumpkins, castor-oil seeds...everything. We must impress the Allocator.' With that Kimeria stood to leave.

Kimeria was impressed with himself. He knew how to handle people. People, especially the useless, poor ones are not that difficult to handle, he thought to himself. With the right words, they could be made to do anything. Didn't they choose him as their Voice without knowing him? And didn't his mother say he was the stupidest person on earth, yet the people of Ukia chose him?

He knew that Wamutheke would be unable to raise anything of substance to move her from the edge of the evil forest...but if she did, well, he would have something more for his store. He watched her sob her eyes dry and touched his eyes too to show that he was also weeping for her. Why wasn't she leaving? He wondered

Wamutheke left with a heavy heart.

XIV

Before Wamutheke went to see the Area Voice, she had to leave Kang'we in a safe place. The ogres had breached her house, twice and devoured half of her beloved grandson. She still did not understand how they had resisted eating the whole of him. It must be God, she concluded.

She still was puzzled about an ogre with a distinct name, Kalondu. This was new. Were ogres evolving and now had names? But how would her Kang'we know that much about ogres, names and all yet she, with all her age and knowledge did not know of this new aspect?

She knew that she could no longer leave Kang'we in the house. She could also not take him with her to the Area Voice. He was too sick and in pain to go all that distance. She also knew that she could not take Kang'we to any of her neighbours. She was already ostracized and any suggestion that she had attracted ogres would drive her stigma even deeper.

Tears of pain and regret began streaming down her wrinkled face. She started sobbing heavily, her chest heaving in great contractions.

She had nobody, yet she was surrounding by so many. She realized that her poverty was the poverty of hatred...that she was poor because of the feelings that people had towards her. They hated her, because she reminded them of their problems and fears and terrors of what could go wrong in a minute. She started wailing loudly, the wail of poverty and need...calling on God to see her and save her and her beloved Kang'we.

Her wailing awoke Kang'we on his bed. He saw her crying, and he began crying too. The cry of the two, an old woman and her reason for living tore, through the thatch in a soulful and mournful combination of pathos and pain. One cried for fear and trepidation and love, the other cried for pain and regret and love.

As she cried, hot teardrops falling on his forehead, Wamutheke removed the bandages she had tied around the gaping hole in the shoulder. She went out to get more herbs to dress the wound.

She decided that she would take Kang'we up the tree in the compound and tie him up there for safety before she left to the Area Voice.

The struggle to climb the tree, fashion a nest for Kangwe and haul him up the tree took the best of three hours. By the time she was done, the sun was already turning the eastern sky, a violet hue. The sun rise would be in an hour or two. She hauled a few sweet potatoes and a gourd of milk. That woud take care of his meal until she got back.

Then she set of to the Area Voice's residence.

It was not long after Wamutheke left that Kang'we heard the stump, stump, stump of on ogre hopping about the compound. It was Kalondu!

A whiff of Kalondu's evil smell wafted through the leaves and the branches and reached Kang'we as he hunkered into the blanket that Wamutheke had fashioned into a carrier and hung from a branch. He was now not only scared of Kalondu, but he also realized that he had to keep quiet, so that the little ogre can go away.

But the stink from the little ogre was too much. Kang'we's nose began to twitch. Oh no! He was going to sneeze...and give away his hiding place to the little ogre.

'Ahhh...aaahhh'...he tried to squeeze his nose, ' aahhh...aaahhh....schchchiiih!'

Kalondu heard the muffled sneeze.

As Wamutheke met Ms Kimeria at the Area Voice's residence, Kalondo was at the foot of the tree, peering up the tree. His sense of smell was much sharper now that he had tasted human blood. He tried to peer with his one yellow eye into the foliage but he could see nothing.

The smell of flesh was so strong in Kalondu's snout that green drool started dripping down his scaly neck in rivulets.

'Where are you...?' Kalondu could not remember Kang'we's name. ' Are you up there?'

Kang'we stuffed the one corner of the blanket into his mouth and resisted another sneeze. He was terrified and his trembling could have fanned a fire. The terror in his heart killed the pain in his shoulder, as beads of sweat erupted onto his face like tiny grains of millet.

But he could resist Kalondu's smell for so long...

The bit of blanket flew off his mouth as another sneeze, even more powerful than the last one erupted out of his scared chest.

'Ahhh....ahhhh....AHSCHOOOO!'

Kalondu was now crazed with anticipation. His tiny ogre mind had forgotten friendship. He could only remember one thing now...the taste of warm blood and the scrunch-scrunch-scrunch of crispy bone...

He hopped around the tree in crazed greed, trying to catch one of the low branches but he was still too short. On and on, he hopped.

Up the tree, Kang'we sat in terror and the situation would have gone on like that until Wamutheke came back...

Then Kalondu's mother, a four meter tall ugly and scaly ogre, hopped into the compound looking for her lost baby...

XV

Wamutheke was making her way back home when she saw an owl, sitting on the low branch of a *Muthanduku* tree staring at her. It gave a mournful hoot and took off, beating its wings fast and furious.

Wamutheke felt the hairs on the nape of her neck rise.

You did not see owls during the day! She thought. It was a bad omen to see an owl during the day and what was worse, it had hooted at her!

What was the meaning of this?

Had something happened at the Area's Voice's homestead after she left? She thought of turning back to Kimeria's residence.

Something must have gone wrong there. Wamutheke thought. *That woman...Madam Kimeria...something might have happened to her. She didn't look healthy at all...*

She sat down to weigh her thoughts on the matter.

In many times in this world, the weight of worry lies on the shoulders of those the world should worry for and about. Those that need help are in most cases the ones that rush to help those that do not need help.

Had Wamutheke known that at that very moment, her beloved was in extreme danger, she would have flown like the Ostrich that had her house on fire. She would not have thought of the greedy Kimeria as needing her worry. She would have run to Kangwe's aid, to die with him if possible.

She got up quickly to rush to Kimeria's home.

At that very moment, Kimeria was leaving the Area Land Allocator's office. He was accompanied by his fat wife and two fellows that pretended to be his body guards but were actually opportunistic influence-peddlers.

'It is sooo hot,' Madam Kimeria complained as she fanned her bleached yellow face with her stubby fingers. ' Can we stop at the nearest centre, I need   three tubs of cold banana mash...'

Kimeria did not hear her. His mind was on the deal that he had just struck with the Land Allocator. They had agreed to evict Wamutheke from her land.

It is not like she needed that land anyway, he thought. *She has said she needs rescue from the evil forest, no?* He looked out at the dirt road and its scraggly edge raced past under his thin legs.

We will build a huge resort there and create a conservancy out of the evil forest...That child Kang'we will most probably die from that rotting wound and Wamutheke will be glad to work in my compound for her meals...

Kimeria thought of his genius. *He was a genius!*

You create opportunities from problems, he told himself. That is what Prof Wharton had taught them at the Oxerter School of Development Studies. People's problems are your golden goose...You eat from those with problems.

That was the first law of predation...no, wouldn't call it that, maybe first law of opportunism. Get the weakest first. Wamutheke had a problem with primitive stories, he was going to create opportunity from her stupid ignorance and her weakness.

That child had just probably got himself bitten by a wild dog or something. Ha! Ogres my foot! Stupid folks these! We'll show them ogres...he thought wryly.

He saw in his mind the entrance to his resort...KIMERIA WILD-LIFE CONSERVANCY...he scratched his bony head with a long finger that ended in a dirty claw for an appropriate catch-phrase to go with the resort's name...

KIMERIA WILDLIFE CONSERVANCY... 'Chew and Swallow At Your Leisure'

No, he thought, *that sounds too personal.*

KIMERIA'S LAND OF THE WILD... 'Let Us Crunch Your Burdens Away'

No, that one does not tie in with the evil of the forest...

KIMERIA LIFE RANCH... ' Where Life Ends And Living Begins'

Kimeria was getting darker and darker and his eyes got redder and redder as he thought about his latest catch. His driver looked at him through the rear-view mirror and thought Kimeria did not look exactly human. The driver felt a cold shiver run down his spine and he involuntarily stepped on the brakes, sending the huge machine screeching, sending gravel and sand across the road...

'You fool,' Kimeria and his wife knocked him over the head. Do you want to kill us?'

'I – I- I am sorry Mhesh...'

Kimeria settled down. His mind was still on his latest deal. He called his construction manager,

'Kiundutho, let us meet at Wamutheke's...Wamutheke...the old crone I told you about...What? No, you fool! What? At the edge of the Evil Forest...Yes, Maiyoni. Exactly! She lives alone, shunned and spurned...No one is there now, at least no one to stop us... I left her at my home and said she should stay there and not be allowed to leave. She is to live with me from now henceforth, that is what I told them...Yes. Only her stupid grandson is there ...Of course...now you get me...The way we did it last time. If he is in the house...Crush everything and clear everything. If he is inside the house...Yes...if he is inside the house, then too bad. Nobody knew he was there...nobody will find anything... crush everything to powder...Make sure you clear everything. Do you hear, no blood, no bones, no roof, no walls...Yes... I am on my way, let us meet there...'

Kimeria switched off the phone and grinned to himself. Things were going smoothly...like soft steak down the throat...

Wamutheke rushed towards Kimeria's home; the same owl flew ahead of her and perched on a low branch. As she approached, it hooted, 'WHOO-WHOO'. Twice. It fixed her a stern gazing stare.

XVI

Wamutheke stopped dead in her tracks. A cold hand of fear took hold of her heart and squeezed it hard. A tremor of terror shook her entire being and she felt as if somebody had taken hold of her covering of skin, leaving her every cell of being exposed to pain, fear, heat cold, suffering and every kind of unkindness that life could throw her way.

From somewhere within her, she felt a wellspring of anger, fury, wrath and fire bubble forth from the pit of her belly.

 It rose from the depths where fear reigned supreme, killing fear and worry in the process. It rose in waves of mad intention, choking her and demanding movement. That was a storm of vengeance, of fateful protection. She knew straight away. She knew she had little time and her beloved Kang'we was in extreme danger.

To say that Wamutheke ran is to misreport what animals, insects, frogs, birds and one or two hunters saw.

A streak through the forest. A flash of dress. A smell of a human too fast gone out of sight to perceive.

Wamutheke flew to her beloved grandson, tears streaking off her face in furtive droplets as they were swung off her face by the air she cut asunder. She flew.

XVII

The huge ogre had spotted Kang'we and the smell of pus and blood from his wound was too much for her. Her green saliva drooled downwards in long slimy straws. Kalondu hopped up and down in excitement beside his mother.

The monster had tried climbing up the trunk, but ogres were not made for climbing. Its two short limbs that stood for arms were not strong enough to lift the ogre up the branches.

But its jaws were strong enough, so much that if she bit at a branch, she could hold up her entire body so that she could wiggle her one huge stump over the branch and climb up. But none of the branches she was biting into were strong enough to carry her weight. So she bit and broke the branch. Hopped and bit another. And broke it.

Hop.

Bite.

Break.

On and on. Soon, all the lower branches were broken and lay around the two ogres. Kang'we was now clearly visible to them, The mother-ogre stopped and surveyed the scene up the tree with her yellow eye. The eye-slit had grown bigger with anticipation of the feast of the little human up there.

Then an idea from somewhere in the ogre's tiny brain emerged. Even the devil has brilliance moments, they say.

She could push her Kalondu up the tree and get him to tear the human free. Then he would fall on the ground and the two would then eat him. She picked Kalondu by her two scaly arms with Kalondu letting out an ogre-yelp at her touch, her claws puncturing his scaly skin and letting out green blood.

The mighty roar of anger and fury that tore across the land, into the evil forest and up into the dry air momentarily deafened everyone. Dry leaves tore off and blew away as bats fell from caves and birds had their feathers blown off.

Wamutheke had landed.

From a distance she saw the two ogres struggling to climb up the tree of her own compound. Ogres in her own compound! How dare they defile her home with their presence.

She saw the torn branches. From her own tree! The ogres had torn branches from her beloved tree, the tree that was planted by her great-great-grandfather! The tree which they had taken care off over many generations!

How dare they!

She saw that they were trying to get at her Kangwe!

HOW DARE THEY!

She felt incinerating heat boil up her blood and her inner being. The red-hot heat shot out of her eyes, her ears and the million and one pores of her skin. She felt the earth push up her up and give her a shot of courage.

To fight for her home. Her tree. He beloved Kang'we. She felt herself grow bigger and bigger, her whole self transformed.

She felt herself floating as she ran at the ogres, her lungs bellowing out all that she had suffered and was suffering, to visit vengeance on the two greedy monsters that had taken her home hostage.

The two ogres did not wait for Wamutheke's fury to reach them.

Kalondu's mother had never heard such a roar. She tried to painfully cover her holes on top of the head that served as ears, but she felt her scaly skin painfully vibrate. Ogres are known to be cowardly and scared of daring humans. The mother-ogre picked Kalondu where he had fallen when she dropped him and very quickly hopped fearfully into the darkness of the evil forest.

Up to today, Wamutheke does not know how she climbed the tree without the lower branches. But she got up to where Kang'we was, untied him and carried him down the tree, Kang'we had by now fallen into unconsciousness. His shoulder was still suppurating.

Wamutheke knew that the ogres would back for her and Kang'we once they settled down and lost their initial shock. She also knew that now they would come in a greater number, as ogres were known to smell greed off each other and so follow the one that had the smell of the green drool on their skin...

Wamutheke gathered a few items from the house and tied the sleeping Kang'we onto her back. She knew of only one place to go to, where she could find the help she needed. *God bless Hon. Kimeria,* she thought.

XVIII

Kimeria, his wife and employees drove into Wamutheke's compound. There was no gate to speak of. The fence was only tree branches thrown around the compound. There, at one corner of the compound was the hut where Wamutheke and her grandson lived.

Pathetic, Kimeria thought. *How could a human live in such a hovel?*

The hut was slanting towards one side,obviously collapsing. The thatch roof looked like a drunk's cap falling off and Kimeria almost laughed, remembering one of his alcoholic teachers at Oxerter College. The door, made of twigs held together with twine looked broken. He knew that sick brat Kangwe was in there, sleeping his sick body off...That reminded him something.

The destroyers should be here flattening up everything. Wamutheke will have to believe that her grandson had disappeared off somewhere in her absence...*His crushed bones, he was sure,would never be found* his eyed glittered maliciously. The boy should never be allowed to live and come later to claim this land.

Kimeria surveyed the compound and the tiny land around it. He knew Wamutheke did her farming elsewhere. He would claim that piece too. He looked at the huge tree about twenty meters from the house. Someone seemed to have been crudely pruning it.

Somebody must have been here before me! He thought. *Some crazy monster must be trying to steal the tree and firewood from the old lady. Is there no sympathy around these parts? Stealing from an old lady her only means for heat...without leadership, I tell you, people can easily turn into barbarians...monsters! Well, too late, I am here.*

His wife joined him. She was onto her third wedge of mashed banana and bottle of soda. The other members of staff stood around surveying the land.

'Darling'. Madam Kimeria says, ' we must build several tree houses inside this forest. I always loved tree houses...'

At that particular moment, they heard the sound of tractors and two huge tractors moved in. One had a back hoe the other had a battering hammer. They were here to flatten everything.

Kimeria was now very excited, seeing that all his plans were coming to fruition in the space of a day. When was the last time he was this lucky? He could not remember. He gathered everybody around him, first to pray and then lay out the work. They held hands and closed their eyes and Madam Kimeria swallowed the last bit of mush as she prepared her voice to pray to high heavens...

That is when the one hundred and one ogres of the evil forest struck...

**

If you go to Maiyoni, where Wamutheke and Kang'we once lived, there are mangled wreckages of huge machines and two massive tractors. Only the torn vehicles that say nothing about where their owners went to. All the vehicles are destroyed beyond salvage. Thieves did away with many parts from the wreckages.

Nobody knows where Kimeria and his wife disappeared to. Nothing has ever been found to show where they could have disappeared to. No body, no shred of clothing or piece of jewellery. All their important staffs also disappeared with them.

A commission of inquiry was set up but the only thing they could come up with was a notice to members of the public to forward information they may have on Kimeria and the disappeared loved ones to a website address provided. The disappearance of Kimeria, his wife and employees is one of the lasting mysteries of the land.

But if you stay late into the night at Wamutheke's compound, it is said, you will hear cruel growls, waning screams and sounds of crunching bones and squishing flesh, you might smell blood in the dust...And the *thump! thump!* thump -thumping of a hundred thumps...

**

Wamutheke and her beloved live at Kimeria's place. The only member of staff left at the home an old maid who used to cook for the family had told Wamutheke and anyone who cared to listen that Kimeria had left instructions that Wamutheke should live in the house unmolested.

As long as she needed to.